To Fred,
all the best
[illegible].
Charles

THE CREATIVE CITY INDEX:

MEASURING THE PULSE OF THE CITY

BY CHARLES LANDRY
& JONATHAN HYAMS

COMEDIA

First published by Comedia in the UK in 2012
Copyright © Charles Landry

All rights reserved
ISBN: 978-1-908777-02-07

Comedia
The Round, Bournes Green
Near Stroud, GL6 7NL, UK

Book design: **www.hillsdesign.co.uk**
All photographs: **Charles Landry**
Special thanks: **Sue Andrew**
Cover photograph: *The Guggenheim Museum in Bilbao is the emblem of the city's physical regeneration.*

This and other Comedia publications are available from:
www.charleslandry.com

Printed on FSC certified paper, using fully sustainable, vegetable oil-based inks, power from 100% renewable resources and waterless printing technology.

Southend, UK: A school interior which has led to dramatic improvements in students' achievements.

CONTENTS

The Gulf: Two young Arab children studying a map of possibilities.

SUMMARY

"... to be creative you a need a challenge that is just about achievable..." Adapted from Howard Gardner's, *The Creating Mind.*

Cities need to know how well they are doing and evaluating cities is becoming a phenomenon. Various measurement systems exist concerned with innovation, city performance, diversity, quality of life, and the creative economy amongst many.

The Creative City Index assesses the creative pulse of places by exploring their urban dynamics, processes and projects. It differs from most indexes by looking at the city as an integrated whole from an insider and outsider perspective through a series of ten broad crosscutting domains.

Cities are only creative if they display a culture, a set of attitudes and a mind-set open to imaginative thinking, visible in all sorts of initiatives. Many cities have creative projects but are not necessarily creative as a whole. Crucially, uncreative places decline and fail, since they do not interrogate their past and present or reassess their resources, prospects and potential for the future.

Wider conditions, such as the level of openness, determine how a place can harness its collective imagination and punch above its weight. Then specific attributes are necessary components, such as good education and skills or research expenditure. Significantly, the rules and incentives regime of local and national governments set the atmosphere of a city, shape its culture and creative climate.

The Index was originally conceived and developed in 2008/9 in collaboration with Bilbao Metropoli 30, one of the city's long-term think tanks with an important role in helping Bilbao reinvent itself, and Bizkaia Xede, the city region's talent attraction agency.

The strategic conversation that emerges by taking part in the Index surprises many participants. So far it has engaged several thousand people across a diversity of sectors, from Finland to Australia, to talk about their city in novel ways. The experience is seen as liberating. It takes them out of their specialisms. It enables them to work through challenging issues and explore opportunities across disciplines. The value of this even outweighs the scoring and benchmarking system of the Index.

OCEAN
HEIDELBERGCEMENT Group
Concret
Thinkin
for a sustainab
MUSTER STATION
Better Life Through Healthy Living
OCEAN

EXPLAINING THE CREATIVE CITY INDEX

We live in a world of measurement. The world is entranced by measuring things: by length, volume, income, size, energy or some other aspect of reality; measurement of wealth and poverty, happiness and depression levels, whole societies, economies and now even cities.

We are mesmerized by rankings: who is at the top and who at the bottom. We love comparisons and these rankings help to place us in relation to others. *The Business of Cities: City Indexes 2011* by Greg Clark summarizes succinctly over 50 of these rankings[1].

We measure to simplify, to break down complexity and to organize our worlds. To measure helps identify and track progress against aims and targets, to identify opportunities to improve, to compare performance against internal and external standards. Measuring helps clarify our strengths and weaknesses to help formulate and guide strategic goals and activities.

The Creative City Index assessment starts with gathering key facts about a city. This provides a baseline, often statistical. It then has two more elements: an internal, subjective, insider perspective; and an external objective assessment. The differing internal judgements, opinions, and group discussions are extremely revealing. The methodology allows for each element of the process to result in a contributing score. Equally the difference between the results of the internal and external assessment are significant, since cities can either over or under-evaluate themselves. This is helpful for the strategic discussion about the overall results.

The ten key indicators of a creative place

We have distilled creative urban dynamics into ten cross-cutting domains, headings or groups of indicators for creativity. In assessing these each participant has to look at their city as a whole. Within each of these domains there are key traits or

Granville Island, Vancouver: A leftover cement plant now surrounded by cultural centres and arts insitutions and a famous food market.

[1]The Business of Cities www.thebusinessofcities.com/

questions indicating creativity:

- Political & public framework
- Distinctiveness, diversity, vitality & expression
- Openness, trust, accessibility & participation
- Entrepreneurship, exploration & innovation
- Strategic leadership, agility & vision
- Talent development & the learning landscape
- Communication, connectivity, networking & media
- The place & place-making
- Liveability & well-being
- Professionalism & effectiveness

Creativity is clearly not the preserve of any single sector ...

We focus on domains because of their broadness and depth. They cover large interconnected fields of interest and knowledge. They enable people to more easily see themselves in their city and to see their work as relevant to whatever the subject matter, such as the economic or social. For instance, the public and political framework of a city is as germane to an educationalist as it is to the social worker or the business community. Equally, how well the communications work is significant to people and organizations in any field. These domains are not necessarily the obvious atomistic categories but are more holistic, drawing together strands that affect all sectors, people, organisations and aspects of city life.

The sectors examined

Creativity is clearly not the preserve of any single sector and it is important to be wide ranging. The sectors specifically assessed include:

- The education and training system: primary, secondary and tertiary education, professional development, lifelong learning
- Industry and business: SMEs and large corporates, the different commercial and industrial sectors, cluster initiatives, representative bodies such as the Chamber of Commerce
- Public administration and public bodies and facilities
- Professionals in the design communities
- Health and social services

Berlin: A workshop discussing socially inclusive economic development.

- Transport and mobility
- The community and voluntary sector: local societies, social action groups
- Culture, arts and gastronomy
- Leisure, sports, the hospitality industry and tourism institutions
- Media and communications industries

A city, like an organization or individual, needs the pre-conditions of creativity.

The qualities measured

Within each of the 10 domains a strong showing is needed in the following qualities: motivation, tenacity, awareness, clarity of communication, broad thinking, inspiration, aspiration, adaptability, dynamism, openness, participation, design awareness, sensory appreciation, professional pride, leadership, and vision. A city – like an individual or an organization – needs many of these traits if it is to be alert, forward focused and alive: these are preconditions of creativity.

Perth: An amusing hoarding on an old building being refitted to encapsulate the spirit of the creative city idea.

Creating your own destiny

For a city, a central purpose of being imaginative is to create and control its own destiny rather than falling victim to circumstances conditioned and forged by others, or outside influence. There are innumerable factors that shape prospects. Traditionally these were largely seen as immutable and tangible, such as location and physical resources. Today, by contrast, urban assets are more intangible and invisible such as the personality and characteristics of its people and organizations. When these qualities are positive and well aligned it is easier to bend situations to your needs and to guide events your way. This requires clarity of purpose, a scenario of where you want to go and a plan of action. Cities that are able to orchestrate this complex dynamic do much better and can then punch above their weight. It is this quality in particular that we assess in the Index.

The fragility of the creative milieu

A creative milieu is a stimulating environment which makes places more attractive. It is fragile as it draws people in and over-stimulates demand and is easily burnt out – too much buzz and always looking for the next thing. Even if burn-out does not occur, prices are often pushed up threatening affordability in a self-defeating circle. Younger people

or newer companies find it increasingly difficult to rent or buy. It is the negative side of gentrification and now a frequent global phenomenon. While the incoming middle-classes and their businesses can be important in turning ideas into reality, they may not bring in the cultural vibrancy that makes places compelling. Space can run out, pushing people to cheaper more peripheral areas. Berlin, for instance, has suddenly 'arrived' and attracted people from around the world, lured by its artistic atmosphere and cheap prices for space. This was due to putting two cities, East and West Berlin, together so creating a very large spatial template. Yet Berlin's competitive prices are already disappearing – the self-defeating circle.

This price dynamic can make places flip from feeling easy, experimental and exciting to becoming more staid and only reachable for those with some wealth. The challenge for cities is to find a balance where differing people with varied income levels and experience can come together. This means a city has to find a way of bending the market to anchor in lower cost uses, either for housing, office space or cultural activities. The public sector can often subsidize incubator units, social housing or spaces for culture and other innovative activities, all in the wider public interest.

DAMA DE COPAS
DAMA DE COPAS

CREATIVE AND UNCREATIVE PLACES

Of course, not every city is vibrant and energetic or stands at the hub of global affairs. Yet every city has a reason for being, a history and a function. There are more ordinary than extra-ordinary places, fulfilling their role from being a port to being a market or commuter town. Cities rise and fall over time, and not many keep their relevance over the very long term.

Many cities start with similar assets. Maybe they have a good location, they are at the crossroads of trade, they are blessed with natural resources, or they are a centre of power, learning or even a cult. Some make the most of their potential, are alert and forward focused, building on their resources and, when necessary, adapting to new conditions and reinventing their purpose afresh. Others do not.

Some cities find themselves in a difficulty outside their control, for example: their resources have run out (cities with a fishing industry); the goods they produce have priced themselves out of the market due to cheaper foreign competition (steel or textile cities); they are less in fashion now (tourist destinations); or new transport links have bypassed them and shifts centrality (a new railway line or airport hub). For example, the rise of Dubai and Madrid has reduced the position of Bangkok and Barcelona. Cities rise and fall; think of Athens, Jerusalem, Tangiers, Bruges, or Liverpool. Is being a tourist attraction enough? Also think of cities which peaked in the industrial era: Baltimore, Cincinnati and the numerous cities in Eastern Europe. Yet all these could benefit from being inventive and sharp. They could redefine their tourism offering, make higher value goods from steel by moving up the value chain, or create a new niche.

The current crisis reminds us, as Einstein suggests, that the thinking that got you to the problems you have, will not be the thinking that gets you to where you want to be. It now becomes more urgent for us to think clearly, to test our assumptions and to stand back and reflect as to what went wrong and how to move forward. To be willing to be more creative can only help in these circumstances.

Lisbon: The Santa Justa lift is an imaginative solution to connecting the lower and higher parts of the city.

WHAT IS AN UNCREATIVE PLACE?

There are many uncreative places and one thing we know is that in time they will fail and decline in spite of any initial intrinsic advantages. Let us imagine these uninspiring places before describing an ideal creative city.

Many cities unfortunately become complacent, considering only the tried and tested, losing energy and maybe just hoping that luck will come their way. Here, existing power structures hold on to their status, they reject ideas or refuse to let new people in. Rather like a company in decline they do not reassess their business model or think of a new vision and purpose. They become somewhat obstinate and focus on past achievements. This pride can hold them back. They stick to their – perhaps archaic – rules without checking if they suit the current environment. There is an inward-looking dominance, and fewer connections are made both internally and with the outside world. People, businesses, public bodies and organizations generally work alone, rarely thinking that perhaps a partnership could provide a faster and better route to problem solving in fact they fear collaboration, seeing it as a threat and a potential for competitors. They are self-referential and less vigilant and alert.

This dampens the entrepreneurial spirit. Over time this becomes a vicious circle. It is difficult to get resources for renewal and this drains the spirit so its talented people want to leave even if they still have some affection for their place of origin. They have given up hope of meeting their ambitions. Potential leaks out. The city has less scope to invest in its young and its educational reputation tends to go down. Infrastructures look tired, standards decline, and even the urban environment is not well kept. There is too little life in the atmosphere.

Somewhere and everywhere: With more people moving into cities they need to be housed, or is it warehoused.

WHAT IS A CREATIVE PLACE?

On the other hand, it is perhaps quite inspiring to describe the atmosphere, look and feel of an ideal or imaginary creative place.

A creative place is somewhere where people can express their talents and potential which are harnessed, exploited and promoted for the common good. Things get done. These talents act as a catalyst and role model to the development and attraction of further talent. It is a place with myriad, high quality learning opportunities, formal and informal, with a forward looking and adaptable curriculum. The physical environment functions well for its inhabitants, it is easy to move around and connect with each other. Its high level urban design inspires, stimulates and generates pride and affection. The architecture, old and new, is well-assembled, and the street pattern is diverse and interesting. Webbed within the ordinary is the occasional extra-ordinary and remarkable. It is an environment in which creators of all kinds are content and motivated to create and where there are outlets and channels to exploit innovations or for the sale of their work. It is a natural market place, where people exchange ideas, develop joint projects, trade their products, or work in its advanced industries. It communicates well both internally and to the external world and its media is responsible and excited by the city's prospects. It offers rich, vibrant experiences through for example gastronomy, the arts, heritage and its natural surroundings, including thriving mainstream and alternative scenes and a healthy network of third spaces. Opportunities abound: the place is welcoming and encouraging. Its dynamism makes it a magnet and so generates critical mass that guarantees longevity.

The public sector has clarity in its perspective and direction, and understands the importance of harnessing the potential of its people. It is lean, clear and focused. Its workings are easy to navigate and it is accessible, open and encourages participation. Public employees here are focused on the job at hand regardless of departmental boundaries. Differences are a natural part of this discussion culture. They are debated, accepted, negotiated

Genova: Nicknamed 'la superba', the 'superb one', once a significant and courageous place – it is trying to reinvent itself.

and resolved without rancour. Its leadership has vision and is strategic yet is grounded in day-to-day reality. It is respected and trusted and recognizes its vital role in continuously identifying new opportunities and future-proofing. The society it rules over has a high degree of cohesion it is inclusive and fosters participation, is relatively open to incomers and to new ideas, even though these can sometimes be uncomfortable – indeed, creative places are often not that cosy and can be somewhat edgy. This society enjoys its status as a creative hub and the physical environment in which it exists. Levels of crime are in general low, the place feels safe and standards of living are relatively high. It is socially alert and seeks to avoid ghettoizing its poorest. Social organizations are active, well-funded and constructive.

Amsterdam: The annual PICNIC event has become a major forum to explore new trends and ideas.

Taipei: The liveability and convenience of the city has been widely acknowledged. Its advanced transport system plays a crucial role.

Industry is innovative and design aware, with a strong focus on new trends, emerging technologies and fledgling sectors, such as developing the green economy and the creative industries. It is well networked and connected and its commitment to research and development is well above average. Cross-fertilization across even the most diverse sectors occurs as a matter of course. The business community is entrepreneurial, has drive and is forward thinking. It understands and utilizes well its natural resources, and it harnesses existing talents while acting as a breeding ground for new skills. Business leaders are respected figures in their community and give something back. The community in turn is proud of their products and the reputation they bring to the place. Good use is made of its effective communications systems including local and international transport, high-speed internet access and connectivity to the world at large.

... this place has something you might call 'civic urbanity'.

You feel that this place has something we might call civic urbanity. Being civic is to be a full citizen, where people are engaged with their city in multiple ways on an on-going basis in order to improve their lives and those of others. It projects a sense that 'me' and 'my city' merge into one. Urbanity reflects a place that has social mobility, where participation is encouraged and where pride in place is reflected in innumerable ways, from generous public gestures ranging from urban design, or enlightening institutions, or lively activity programmes.

The best of this place comes together in gathering places or civic precincts where public facilities such as a museum, a square or a sidewalk intermesh with private facilities or shops, but where the overall tone is one of civic generosity – the city giving back to its citizens. These desirable places enable us to experience the moment: a place open for coincidence – rather than having to do something specific and always move on to the next thing.

Frieze Art Fair, London: A visitor reflecting on a Russian artwork about bureaucracy.

Auckland: The town hall today is more open and transparent.

Overall, as in all creative places, this place is unlike any other. You can feel and sense the buzz, it is obvious to residents and visitors alike. It accentuates its distinctiveness in a relaxed and unthreatening way. It is at ease with itself. Its history, culture and traditions are alive, receptive to influence and change, absorbing new ideas which in turn evolve and develop its distinctiveness and culture.

... at ease with itself and receptive to influence and change.

Each of the ten domains has its characteristics:

1. The political & public framework

This domain refers to the public institutions, to political life, to government and governance, and to public administration. In an ideal creative place these institutions will be lean but proactive, ethical, transparent, accessible and enabling. Structures will be horizontal and co-operative and departmental lines thinly drawn. Bureaucracy is kept to a minimum. Personnel in the public sector are highly motivated and there are strong links with the private

Seattle: The city has emerged as a global innovation hub. Here, the EMP, financed by Paul Allen of Microsoft.

sector. A healthy community and voluntary sector are encouraged and the general attitude of politicians and officials is to be enabling.

2. Distinctiveness, diversity, vitality & expression

... there is a clear identity and dynamism.

In a creative place, there is a clear identity and dynamism. Citizens are self-confident and proud but open at the same time, inclusive and receptive to outsiders and outside influence. They feel at ease in their city. The cultural offering is wide and welcomes debate and critical thinking. The arts are dynamic and high quality as well as experimental and ground-breaking. Heritage, gastronomy, attractions, parks and the natural environment all add to the vigour of the place. Here is a design-aware environment in which the creative industries flourish, where there are many independent shops, the chain-culture is in the minority, and the retailing experience attractive and special. Expression and debate are encouraged.

3. Openness, trust, accessibility & participation

To be creative, the place needs to be open minded and welcoming and as a result many people from a diversity of backgrounds make it their home. Openness pervades the way society, institutions and organizations operate, creating an enabling environment where opportunities are facilitated and it is easier to get projects going. This attitude is echoed in the inviting way facilities work. It is also a well-connected gateway to and from the world. Its intercultural approach focuses on what people share across boundaries, recognizing difference but seeking out similarities. It downplays ideas of purity and encourages bridge-builders. It acknowledges conflicts and tries to embrace, manage and negotiate a way through them based on an agreed set of guidelines of how to live together in our diversity and difference. It is on the way to achieving the pluralist transformation of its public space, institutions and its civic culture.

Womad, UK: Festivals provide good opportunities for people of difference to come together.

Istanbul: Selling all manner of hubcaps in an entrepreneurial way.

4. Entrepreneurship, exploration & innovation

This place is one where entrepreneurs feel very much at home, where an idea can become reality quite quickly. It is a place where you can make mistakes without being too severely judged. There are extensive support systems from advice to access to funding and risk capital. Clusters, where appropriate, are encouraged to help force-feed innovation and generate critical mass. Rewards, prizes and other recognition systems celebrate achievement and thus there is a higher than average level of innovation and R&D. Universities are keen to turn their insights and research ideas into useful products and services. The open innovation ethos based on sharing and connecting small enterprises and large corporates is well developed. The creative industries play a significant role and there is a reputation for design-led distinctive products and services. Going green is seen as a catalyst to create innumerable innovations.

... you can make mistakes without being too severely judged.

5. Strategic leadership, agility & vision

In a creative place, there are dynamic and forward-looking people of quality in every sector providing a strong sense of vision for the place. New trends and emerging developments are flagged early (currently, the green agenda would be a perfect example). Leadership style is noticeably inspiring, able to delegate and be empowering to others. Thinking is strategic and future-proofing. The decision-making communities in public and private walks of life have a forward focus, whether they are teachers, public servants, transports chiefs, middle and higher management in industry and business, or community organisers or those in the artistic world. There are good mechanisms to bring people together from different disciplines as well as gather information on best practices and innovative solutions from around the globe.

Sao Paulo: Lack of job opportunities makes people create inventive public services.

Kirovograd, Ukraine: A group of fresh-faced girls from an academically rigorous school that has a strong arts programme.

6. Talent development & the learning landscape

Here, learning and knowledge are highly valued. All talents are nurtured, fostered, promoted, rewarded and celebrated. There is a diversity of learning options with ladders of opportunity that take people up the levels. People of all ages enjoy the challenge of learning and want to self-improve. Schools connect with the local community in multiple ways and share their facilities. Universities identify with and are committed to the city, as in Portland, Oregon, where the slogan 'Let Knowledge Serve The City', is emblazoned across the entrance. They open themselves out and contribute to helping solve urban issues. There is teaching of core skills as well as centres of excellence that are globally recognized. There is pride in teaching and the education institutions strive to be the best in their field. The constantly evolving curriculum is in tune with the needs of business. This system grows and retains talent, and there is also a two-way flow which places that talent abroad when appropriate and brings other abilities in as needed.

7. Communication, connectivity & networking

A creative place is well connected internally and externally, physically and virtually. It is easy to get around and ghettos are rare. Social mobility is more possible: diverse cultures connect. There are high quality public transport systems. It has a sophisticated IT and communications infrastructure. The population travels at home and abroad taking advantage of the excellent rail and air services which also make a gateway for receiving outsiders. Speaking foreign languages is common place. Business to business and cross-sectoral links work well, there are clusters, hubs, focal points and knowledge exchanges. The place is outward looking and makes contact at all levels abroad, creating joint ventures, research projects, product development and civic partnerships.

Bilbao: Getting ready for a night out, to communicate, network and possibly much more.

Dubai has massively transformed – at times garishly and at times quite well.

8. The place & place-making

A creative place uses its collective skills, techniques and insights to make this place special. Its urban design teams orchestrate and weave its elements together involving planners, highway engineers or developers collaborating with others who understand how the social, cultural and economic life of the city works as well as those who think artistically. The built environment is human centric and sensitively conceived and implemented. Human interaction and activity is encouraged by this physical environment rather than being blocked by physical barriers (uncrossable main roads, railway lines or closed private blocks). It acknowledges and

respects and blends well with its natural environment, its surrounding landscape, and its green areas, and is aware and responsible regarding its ecological footprint.

The public realm acts as the connective tissue within which the buildings, forecourts and streets form a pattern or mosaic. The urban design knits the parts of the city together into a more seamless whole so each element gains from its proximity to the next. When you are there you want to be there but its reputation drew you there in the first place – it has a critical mass and a magnetism which enables it to compete well with other places which have similar mass and attraction.

Portland, Oregon: A popular city beach in downtown.

9. Liveability & well-being

... there are many opportunities to become engaged.

You will find an exceptional quality of life here. GDP is high and services work well and are of a high standard. People are generally happy to live or work here, appreciating the low levels of crime and violence and feeling generally safe. Health, housing and social facilities are well provided and well run. There is a good atmosphere and people help each other more willingly. Community life is active and there are many opportunities to become engaged and to participate. It is an inclusive place. People gladly take responsibility for their environment. There are good housing choices for all ages and varying circumstances, with rising house prices kept in check. The civic leadership is mostly respected and trusted.

Berlin: Relaxing by the Spree River.

10. Professionalism & effectiveness

The creative place works well, things happen and are achieved. There is pride in being professional and doing things with quality. Standards are high and benchmarks are frequently set here. Companies, organisations, individuals and products are often given awards. This is a centre of expertise in a range of specific areas – attributes such as reliability, punctuality, efficiency or accuracy are highly respected. Professionals are confident in their own ability and not afraid to work in partnership with others and to delegate authority, breaking with conventional rules of hierarchy.

JAMISON
SQUARE

instalações com
ligação à central
de alarmes.
CHARON
SANTA
CASA
CRECHE
NOSSA SENHORA DA CONCEIÇÃO

EXPLORING THE CREATIVE CITY INDEX

Vitality & viability

Cities are living organisms. They have periods of growth, stasis and decline. The chief purpose of acting creatively in an urban policy-making setting is to encourage cities to become more collectively energetic and vital and ultimately viable and vibrant. Assessing this capacity is a central aim of the Creative City Index. Being viable involves long-term self-sufficiency, sustainability, adaptability, flexibility, the capacity to change, self-regeneration, security and resilience. Creativity helps urban viability as it increases the ability to respond to changing circumstances. This is more successful in places whose economic, social, cultural and environmental dimensions are evenly balanced in a sustaining way.

... being creative is the catalyst and process to make the most of resources.

How does being creatively vital connect to becoming viable and resilient in the longer term? Vitality is the mass of activities happening in a city, it is the raw material. These activities need direction, aims, targets and ethical and political goals, otherwise they are merely self-expression. Vitality needs to be harnessed to encourage viability. Here, being creative is the catalyst and the process which tries to make the most of resources.

Critical mass

Critical mass is a crucial concept in assessing the potential impact of creativity in a place. It is concerned with achieving appropriate thresholds which allow activity to take off, reinforce itself and cluster.

In economic terms critical mass involves having sufficient activities to ensure that economies of scale, inter-firm cooperation and synergies can take place, such as having a financial district in the city centre, an artisan quarter in an inner area, a science park on the edge of town or a managed workspace in an outer housing estate. There are minimum thresholds for organizing economic projects such as trade fairs or complex economic initiatives.

Lisbon: Two older women exploring the local buzz.

Socially, critical mass is the number of interactions within the various parts of the city at different times of the day. In environmental terms it could be the density of historic buildings or green spaces. In cultural terms it is its history, its image and its activities. Can you, for instance, experience in the same evening a French bistro, a play, a late night cabaret in a wine bar and then enjoy a stroll through a pleasant historic area or by contrast meet up with friends, go to a football match and then end up playing snooker?

Critical mass generates the necessary momentum to help initiatives take off and develop full potential. Where it is insufficient, the task is to find imaginative ways to make more out of less. For example, co-branding, redefining and regrouping smaller, sometimes isolated, underrated and underexploited existing strengths, where the totality can become more than the sum of the parts.

Holistic criteria

The city is a complex entity, it needs to be assessed from a cross-cutting perspective.

A city is a complex and multi-faceted entity. We can see it as an industrial/commercial/financial structure (an economy); a community of people (a society); a designed environment (an artefact); a natural environment (an ecosystem); and all are governed by an agreed set of political rules (a polity). The ten domains for assessment in the Index cut across all these four dimensions of what constitutes a 'city'. They can make sense from different perspectives or viewpoints such as those of varying professional backgrounds or disciplines. These include politicians, academics from diverse fields, planners, architects, entrepreneurs, city centre managers, the police, social activists and ordinary citizens.

Each domain is intrinsically cross-cutting. It should be assessed holistically and can be relevant to any topic. For instance, distinctiveness, communication, or learning can be relevant to social services, business, cultural life or physical planning. Equally, openness or professionalism impacts upon the social, the cultural, the environmental, the economic and the political arenas. Similarly, social cohesion implies openness or communication and social mobility between groups. By contrast the overall culture of a place could be closed-minded, which would have implications

Tel Aviv's White City is now a World Heritage site for its, 'outstanding example of town planning'.

Bedford Park, London: Based on the garden city ideal, it reminds us that good suburbs are part of the creative city.

in terms of economic life. Entrepreneurship is not only an economic concept concerned with the private sector and business issues. The public sector can be entrepreneurial (within accountability principles) and social enterprise and social innovations are crucial in creating balanced development in a city.

Considerations & assumptions

The Creative City Index has a number of distinctive features. Most importantly it looks at the city as an integrated whole where many dynamics interweave capable of either reinforcing or counteracting each other.

The Index examines the creative processes going on in a city, looking for a deep culture of creativity, of imagination, of thinking, and across a wide spectrum of activities. The presence alone of creative projects does not make a creative place. A city might have a vibrant new media industry yet very traditional transport or environment departments that may hinder clustering in that sector, or it might operate in a top down way, so

narrowing the scope of imaginative action by empowered citizens.

It is necessary to stress again that the Index seeks to be clear about the difference between creativity and innovation. They are related, but crucially they are not the same. Indeed in commissioning the original Index Bilbao said something very brave: "we know we are innovative, but we are not sure how creative we are". They know they are very good at implementing existing inventions and innovations in fields as diverse as the automotive industries and wind turbine technology. Yet this is not the same as inventing something afresh.

A creative atmosphere in any sphere, be it the political, scientific or social, requires, as a prerequisite, both curiosity and an exploratory mind. From curiosity can stem being imaginative, and from this creativity can emerge. These first steps are divergent. Then a convergent process starts. Any idea or proposition needs to go through a reality check from which prototype inventions can emerge, and which if generally applied can become innovations. These in turn foster curiosity and so the cycle begins again.

A creative city attempts to develop and sustain a framework which supports infrastructurally the hardware and software - and the orgware – required for imagination and creative behaviour and values. So this creative climate, as it were, enables innovation, partly by ensuring fundamental prerequisites like excellent healthcare, social services, good education and skills training, or research expenditure.

Size, location & context

Cities come in all shapes and sizes, from the small market town to the megalopolis ... each can be creative in their own way.

Cities come in all shapes and sizes, from the small market town to the megalopolis. The question then is, how is it possible to assess creativity in all these places when the differences are so great, physically as well as culturally. In practice, it is better and more useful to consider like-for-like especially when making comparisons, but some principles are constant, for example the need for openness. An urban typology is helpful. There are different tiers of cities, perhaps as many as ten. At the top are the global centres like New York and London where political, economic and cultural power agglomerate and global agendas are shaped. Then a second group of powerful places such as Sydney, Shanghai and Singapore, are very strong in a cluster of industrial sectors that have global influence. A third level includes places like Barcelona and Milan, with very strong specialisations and niches (such as fashion and design). And so right down to the last level, it would probably be the small market town.

One crucial point in assessing the capacity of a city is determining whether it is punching above its weight and thereby defying expectations. This may put a city onto the same tier as a higher level city. Of course one would not wish to make a direct comparison of, say, Verona with Shanghai.

There are natural assumptions about what different types of city should or could achieve. A capital city, like Beijing, Helsinki or Kuala Lumpur, will typically accrue resources: It will be the communications, commercial and educational hub, headquarters will be located there, it will have a higher proportion of representational buildings, or it will be the centre of its national

Stanford University, Silicon Valley: An exhibition, where the internet revolution started. The region's urban design is not inspiring, but innovation thrives.

Revolutionary Tides

culture. A regional capital is more likely to have industrial strengths and other assets, such as Lodz in Poland or Gothenburg in Sweden. Its issue will often be whether it can draw power from the primary city and establish specialisms not present in the centre. A heritage city typically rests on its laurels. Often it becomes a tourism destination open merely to the tourist gaze, perhaps even mothballed like an old fusty museum. It can retrench into its past successes without fostering much new life. If it were to try to be creative, this city would confound our views – as has Oulu, for other reasons. Oulu, a relatively isolated city high up in northern Finland was able to achieve substantial research strengths and develop innovations like the mobile phone. Equally, external political constraints can also limit potential: Penang is attempting to be forward-looking, strategic and imaginative, but since it is run by the opposition party of the government it gets less opportunities than other places in Malaysia.

Oulu, Finland: Renowned for its extensive technopolis, here is the theatre and library complex.

The methodology of the Index allows all these factors to be assessed.

Methodology & approach

The Creative City Index uses as a statistical baseline the key facts gathered about each city: levels of employment, activity within different industrial sectors, voting patterns and participation rates, cultural and recreational facilities, amongst many others.

The internal and the external assessment combine an insider perspective through the eyes of the citizens and a more objective outsider view. First a wide ranging series of one-to-one and group interviews with knowledgeable and credible people across a diversity of sectors is undertaken, where interviewees individually and collectively

OPEN
30
29
28
27

assess how well their city is doing, scoring each domain on a percentage basis. Separately, web-based questionnaires are completed to capture a far wider audience. All samples take into account different backgrounds and social groups. The results resolve to a series of scores based on weighting devices and varied analyses, allowing them to be viewed in a range of ways and compared with other participating cities.

One feature of the internal assessment is the group meetings, bringing diverse people and interests together to discuss a joint agenda from differing experiences. Here no individual's knowledge or discipline is more important than another's when discussing the domains. This often provides a rare cross-disciplinary opportunity to discuss the city and is much appreciated.

... doing the Index provides a rare cross-disciplinary opportunity to discuss the city.

The external evaluation is carried out by those undertaking the research and interviews, who need comparative knowledge about the creativity of cities and global urban dynamics. Comparing the internal judgements, the group evaluations, and the internal and external results can be extremely revealing.

Currently 20 cities are part of the Index and the aim is that they use the system as a benchmarking tool and repeat the self-assessment and external analysis at regular intervals, for example biennally.

Lessons & insights

Assessing the Index in nearly 20 cities (at the time of writing), a number of lessons have emerged. The process can have significant impact because it forces a strategic conversation across sectors and disciplines, it questions received discipline-based wisdom and brings up issues that would not emerge if things were always examined in a siloed way: it encourages collaboration. It fosters a mutual learning process and it reframes debates about the future of the city. Its comparative nature encourages best practice to be spread between places.

No index is perfect and we have tried to be practical and in particular we stress the combination between subjective and more objective perspectives. Indeed many indexes which claim to be completely objective by providing a specific number or percentage, in reality often draw on some of the subjective views of those putting it together.

Sao Paulo: Galeria Florida in rua Augusta is part of a creative quarter, with many younger start-ups.

Crucially, we take heed of the cautionary words of Daniel Yankelovich the renowned American pollster who noted:

"The first step is to measure whatever can be easily measured. This is okay as far as it goes. The second step is to disregard that which can't be measured or give it an arbitrary value. This is artificial and misleading. The third step is to presume what can't be measured isn't really important. This is blindness. The fourth step is to say that what can't be easily measured really doesn't exist. This is suicide!"

... our view of cities is a mix of reality and truth, hype, image and perception.

The general view of cities today is a mix of reality and truth, hype, image and perception, usually filtered through media representations. This affects the potential of a place: Can it keep or attract skills and expertise, or persuade interesting companies to relocate – drawing power is important in creating magnetism. The right magnetic blend makes a city attractive and desirable, the different aspects tempting different audiences: power brokers, investors, industrialists, shoppers, tourists, knowledge nomads,

Sao Paulo: Hidden in the favelas there are many inventive initiatives.

Learning environments like this do not encourage curiousity and imagination.

the cultural cognoscenti, property developers, thought leaders.

The diversity of views on a similar topic can be astonishing despite the fact that people are looking at the same reality. In Canberra, some thought the political framework was extremely good and gave it a score of 70 percent, whereas others gave it only 10 percent. In Perth some considered the place making to be of high quality (80 percent) whereas others said it was very low (20 percent). In terms of strategic agility people in Freiburg expressed a range between 20 and 70 percent. In Ghent, the communications capacity ranged from 25 to 85. In Oulu, many felt that there was a high level of acceptance and open-mindedness, whereas others felt it masked an inward-looking culture. In Bilbao, there were differences in assessing the entrepreneurial spirit.

If the scores are averaged, say 53 percent, and taken in isolation this might imply that things are alright, whereas they may cover up intensely held and contrary views. The strategic conversations leading to the score are often more significant than the score itself. The wide ranging views expressed provide the city with material to work with and change.

An over-riding issue that has emerged is the silo thinking and lack of collaborative discussion between sectors. Despite years of stressing the need for joined-up thinking,

Copper River Salmon!
MAY 17
MAY 17
PLACE ORDERS NOW!!
FISH
smoked salmon &
fresh seafood worldwide!
Fresh
Daily
ASK US!
FRESH WHOLE STURGEON
12.99
FREE - HOTEL DELIVERY
FANCY JUMBO GRILLING SCAMPI'S
FANCY ALASKAN KING CRAB SECTIONS
FROM THE COLD CLEAR WATERS OF ALASKA
WE CLEAN
CRABS
WE PACK FISH TO GO FOR 48 HOURS
FRESH MEDIUM RED KING SALMON
FRESH JUMBO GRILLING SCALLOPS
FRESH MONK FISH
$14.99
POUND
FRESH HALIBUT CHEEKS
COOKED PEELED DEVEIN SHRIMP
21.50

cities are a long way from working through issues or potential together. Many now work in a multi-disciplinary way where each profession or area of expertise provides their input, but they do not in the discussion reassess their own assumptions. Very rarely do they work in a trans- or inter-disciplinary way, where open conversation and debate cross-fertilizes differing insights and changes the way issues are looked at.

Subjective opinions seen in unison can reveal objective truths. It is interesting how reasonably well-informed people, very quickly come to similar conclusions. This is the 'wisdom of crowds' argument. In Bilbao a wide range of audiences were interviewed over several months. The overall score was 62.25. One reason for this strong score was the city's ability to achieve things in spite of its internal 'war'. In a three hour meeting with around 100 people the Index was explained and discussed and the audience was asked to score the city. Almost incredibly, the result came out at 62.15. Similar events have occurred elsewhere.

... an over-riding issue is silo thinking and lack of collaborative discussion.

A number of methodological issues remain including: the availability of data sources to provide the baseline facts; their degree of comparability and relevance between cities of different sizes; the proportion and extent of attributes measured; the specificity of the local context; the dynamic nature of cities (as post-hoc statistics cannot tell us about hopes, aspirations and goals); the appropriate weighting; and the differences between quantitative and qualitative data. Not everything can be determined objectively and there are four different ways to deal with this:

- subjective measures of subjective phenomena (how safe do people feel?)
- objective measures of subjective phenomena (how much do people spend weekly on taxis because they are afraid to walk home at night?)
- subjective measures of objective phenomena (are people satisfied with street lighting in their neighbourhood or the frequency of public transport?)
- objective measures of objective phenomena (how many companies are working in the creative industries sector in a city or how frequent is the bus service?)

Seattle's technological strengths have spilled over into its food culture.

Of course, objective data can be quantified and measured, while subjective data can only be assessed and judged. In looking at something as complex as a city it is unlikely that a simple set of quantitative data will provide an accurate picture of its lived experience.

It is also important to understand the fluid and cross-cutting nature of the domains we are using. While some subject specialists like educationalists, architects or urban designers might assume that their sphere is only covered within one domain, this is not the case. All the domains have relevance for them. For example, the learning system in a city can be open or closed, it can connect to its surrounding communities or operate as an island onto itself. It can be entrepreneurial or very set in its ways, the political framework can foster new approaches or hold things back. Equally design and planning can have a good public framework, which ultimately determines what is built, buildings or urban design can be distinctive or bland, and can foster good liveability and well-being.

Results & consequences

When it comes to using the results, some cities are simply pleased to have a benchmark against which to judge future progress, and some also like the comparison. Most cities evaluated so far have used the Index for a specific purpose or responded to the results. For instance, Canberra received a relatively good score overall of 54 percent, but was pulled down by an internal score for the city of 34 percent for strategic leadership. This has now become a significant issue also taken up by the media. Canberra, it appears, is good at making plans, but less good at making them happen. So a project team was created, as a result, to reassess how the city's regulations and incentives structure can be changed to help the city 'walk the talk'. In Ghent the purpose of doing the Index was twofold: Ghent already had an imaginative creative city programme and wanted to evaluate how well it was doing; and there was a strong aim to get the city to think and work together in a cross-departmental way. Participating in the Index provided a framework within which to hold that conversation. In Oulu their aim was to embed creativity more into the public administration and the result was an urban culture strategy that focused on looser regulations, raising the profile of the cultural department and a greater focus on the public realm.

Federation Square, Melbourne: A city that experiments with urban design and always scores highly on global liveability rankings.

SBS

ENTRANCED BY MEASUREMENT

In the last decade the ranking of cities has become a new fashion – even a frenzy – as they try to project themselves onto the global radar screen. A few indexes look at the overall performance of a city, others a specific aspect, such as: its business or economic environment or investment climate, its quality of life, liveability or well-being, its sustainability, its image and brand, its level of opportunity, its cultural life, degrees of attractiveness, how happy people are.

Rankings and league tables have become competitive tools. A good city ranking in some field, such as being liveable, economically competitive, or green, is increasingly used as a lever and marketing device.

New concepts are continually emerging such as being a 'smart city', 'innovative city', 'healthy city' and here a 'creative city', and there are indexes to measure each of these. Each taps an important dimension of city life and if taken seriously implies dramatic changes in city development. To be comprehensively 'green' requires reorganizing an urban economy and behaviour change. To be 'creative' demands changes in mindset and perspective. To be 'smart' demands the use of new technologies in innovative ways. To be 'healthy' demands a changed physical pattern of urban development.

All ranking systems have to be looked at with caution. A core question is always, who is commissioning and organizing the study. Also cities are often assessed from an expatriate perspective (Mercers, Economist Intelligence Unit) and with the limiting corporate or world view of multinational companies. Local distinctiveness is rarely in focus. Serious cultural misinterpretations can colour the assessments. For instance, safety and perception of safety is a key to a liveable city. Yet an outsider who cannot read the codes of a city like a resident can mistakenly perceive a city to be unsafe.

Some indexes are devised by specific cities, such as that of the Singapore Centre for Liveable Cities where the city scores very high in its own rankings. Indeed the Creative City Index carries

Moscow: Mokhovaya Ulitsa, near the Kremilin. A symphony of symmetries and proportions.

Vancouver is widely acknowledged to have a strong planning regime in place that has helped its success.

within it the assumption that to be curious, imaginative and creative fosters an urban environment that over time develops a more resilient city able to future-proof itself. This is because it is alert to changing circumstances and is able to generate the preconditions for a place to think, plan and act with imagination by finding the solutions and opportunities to move forward.

... the overall aim is to develop a more resilient city able to future-proof itself.

Two broad trends are noteworthy: who is on top; and who dominates special areas or niches such as innovation, liveability, sustainability or fashion and design which bring in a completely different group of cities. Within the second set, Zurich, Vancouver Vienna, Munich and Nordic cities appear most frequently. In all of

these ranking systems European, Australasian and Canadian cities generally do well. This has led some commentators to suggest that there is a cultural bias to the rankings with Anglo-Saxon places scoring more highly than they should. Definitions, for instance, as to what constitutes liveability, can be culturally specific. Here the annual quality of life survey by the magazine *Monocle*[1] is perhaps the most sophisticated.

... some say there is a cultural bias to rankings.

There is, therefore, a battle as to how to measure the significance of places.

[1]www.monocle.com/webprogrammes/Quality-of-Life-Index/

RETHINKING THE MEASUREMENT OF SUCCESS

The shifts in city development from industrial to post-industrial ones have changed how we measure the success of a place.

Classical measurements of city success	Newer measurements of city success
• Location	• Economic profile
• Physical characteristics	• Market prospects
• Infrastructure	• Tax levels
• Human resources	• Regulatory framework
• Finance & capital	• Labour climate
• Knowledge & technology	• Suppliers and know-how
• Industrial structure	• Utilities
• Institutional capacity	• Incentives
• Business culture	• Quality of life
	• Logistics
	• Sites
	• Community identity & image

The 'Newer' measurements largely overlap with the 'Classical' ones, but quality of life questions and issues of identity and image become more significant.

Sri Lanka: Colombo environs – an unusual street scene.

All these measurements are still largely valid. However, now the distinction between 'old economy' (industrial era) and 'new economy' (knowledge based) thinking is relevant. The main considerations in the former were cost and scale. In the latter, the production cost dimension of competitiveness is still important, but additional considerations, such as quality, talent, innovation, connectivity and distinctiveness come into play.

... Every period in history has its own measurement priorities.

New features emerge such as community identity and image, softer qualities which are less easy to measure.

The drawing power index (developed by Charles Landry with Franco Bianchini, in the mid-1990s) is based on the newer thinking about what makes cities work, and looks at the city from the inside and outside, both through calculable and countable realities and perceptions and reputations. Its assessment criteria are:

- Critical mass
- Identity & distinctiveness
- Innovative capacity
- Diversity & accessibility
- Security & safety
- Linkage & synergy
- Competitiveness
- Organizational capacity
- Leadership

Even more recently, new criteria are being acknowledged as important in driving urban success. Key notions include:

- Understanding the use of iconic communication
- Embedding design consciousness
- Eco-awareness
- Recognition of the power of artistic thinking
- The level of cultural depth in a place

New York: A view from the very successful High Line, established through pressure from a citizen's campaign.

- The quality of its overall atmospherics & experience
- The associational richness & resonance created
- Networking capacity of the city
- Communication & language skills

These newer criteria underpin much of the underlying thinking and methodology of the Creative City Index.

STEP
BY
STEP

A STRATEGIC CONVERSATION

The experience of being involved in the Index has been positive for cities. It has engaged several thousand people who in general have found it to be a stimulating process – because it engages people across sectors to talk about their city in a novel way.

Taking part provides a rare opportunity to be released from professional constraints and predominant ways of thinking, and leads to a discussion of issues of common interest. In fact, the value of this even outweighs the scoring and benchmarking system that results from the process, though this too of course offers very widespread benefits and applications. However, this strategic conversation surprises many participants and almost in consequence moves the city agenda more rapidly forwards.

Participating in the Indexing process is in itself a creative activity which can give rise to innovations in some areas later on. It enables people from different backgrounds and interests to discuss the potential and problems of their city from a broad perspective. They find they can see more clearly how the dynamics hang together. Hence an important cross-sectoral conversation starts, empowering both those already with power and influence as well as those not normally involved in urban strategy. Participants can – and feel they can – make a contribution and create impact, and this generates excitement about the prospects, the future trajectory, the ambitions and the potential story of their city. This is particularly valuable in public service – if public officials are committed, motivated and engaged, results are more quickly visible and the effect can be infectious.

Berlin: Interesting graffiti at the Berlin Wall.

數位迪化設計展
Digital Di-Hua Design

BIBLIOGRAPHY

Andersson, David E., Andersson, Åke E., Mellander, Charlotta (2011). *Handbook Of Creative Cities*. Elgar Publishing

Clark, Greg *The Business of Cities* www.thebusinessofcities.com

Cooke, Phillip., & Schwartz eds. (2007). *Creative Regions: Technology, Culture and Knowledge Entrepreneurship*. Routledge

Csikszentmihalyi, Mihaly (1996). *Creativity: Flow and the Psychology and Discovery of Invention*. New York: Harper Perennial

Economist Intelligence Unit's Quality-of-Life Index www.economist.com/media/pdf/QUALITY_OF_LIFE.pdf

Florida, Richard & Tinagli, Irene (2004). *Europe in the Creative Age*. Demos

Hall, Peter (1998). *Cities in Civilization: Culture, Technology, and Urban Order*. London: Weidenfeld & Nicolson

Kaufman, James C., Sternberg, Robert J., eds. (2006). *The International Handbook of Creativity.* Cambridge University Press

Hollanders, Hugo & Cruysen, Anne (2009). *Design, Creativity and Innovation: A Scoreboard Approach*. Innometrics

Kao, John (1996). *Jamming: The Art and Discipline of Business Creativity*. Harper Collins

Landry, Charles (2007). *Lineages of the Creative City*, download from www.charleslandry.com

Mercer, Quality of Living survey www.mercer.com/articles/quality-of-living-survey-report-2011

Milbers and Sirilli (2008). in *'Wider conditions for Innovation'* NESTA paper by Diane Coyle and Mary Beth Childs, Enlightenment Economics,

Monocle: Quality of Life Index www.monocle.com/webprogrammes/Quality-of-Life-Index/

Moretti, Enrico (2012). *The New Geography of Jobs*. Houghton Mifflin Harcourt

Paulus, Paul B., & Nijstad, Bernard Arjan (2003). *Group creativity: innovation through collaboration*. Oxford University Press

Taipei has an urban acupuncture programme which creates interventions to act as catalysts for regeneration.

STUDIO 1

CHARLES LANDRY

Charles Landry gives tailor-made talks on a wide range of topics. He undertakes task-specific projects or research, and residencies from one week to three months, where a specific challenge or opportunity is worked through and presented publicly at the conclusion. Other projects involve deeper, longer-term relationships with cities and organizations, where strategies are created and tracked, reports are written and meetings chaired and facilitated.

JONATHAN HYAMS

Jonathan Hyams has developed innovative software for the cultural sector and is a director of Artlook Software. He chairs The Conversation UK, a new media company concerned with unlocking university research and expertise to the public, and he has worked with Charles Landry and Comedia for twenty years. The Creative City Index uses bespoke software and analysis tools developed under Jonathan's direction.

Contact:
To book Charles as a speaker, to discuss a project and to enquire about The Creative City Index go to:
www.charleslandry.com

Books:
The following titles can be purchased online and from bookshops. Comedia Shorts are also available as e-books. Discounts are offered on bulk orders over 10 books, for events, workshops and meetings.
www.charleslandry.com.

Titles by Charles Landry:

Comedia Shorts 01: *The Origins & Futures of the Creative City*.
ISBN: 978-1-908777-00-3

Comedia Shorts 02: *The Sensory Landscape of Cities*.
ISBN: 978-1-908777-01-0

Comedia Shorts 03: *The Creative City Index*.
ISBN: 978-1-9087770-02-7

The Creative City: A Toolkit for Urban Innovators, Second Edition. Earthscan
ISBN: 978-1-84407-598-0

The Art of City-Making. Earthscan
ISBN: 978-1-84407-245-3

The Intercultural City: Planning for Diversity Advantage. Earthscan
ISBN: 978-1844074365

Ghent is an imaginative city with a determined creativity strategy. Here, a flash mob event protesting against the treatment of women.

'Taking part in the Creative City Index has really helped our city move forward in an imaginative way.'

Daniel Termont, Mayor of Ghent